Name _____

Chapter 1 Test

Mark the ○ for your answer.

| 0 | | 2 | | 4 | 5 | 6 | | 8 | 9 | | 11 | 12 |

1. 5 is just before ____.

   3 ○   4 ○   5 ○   6 ○

2. 12 is just after ____.

   8 ○   9 ○   10 ○   11 ○

3. What comes between 7 and 9?

   6 ○   7 ○   8 ○   10 ○

4. How many hops from the to the ?

   1 ○   2 ○   3 ○   4 ○

5. How many hops from the to the 🌸 ?

   3 ○   4 ○   5 ○   6 ○

6. 10 is 3 more than ____.

   7 ○   8 ○   9 ○   12 ○

7. What comes next?

   11, 10, 9, 8, ____

   7 ○   8 ○   9 ○   10 ○

8. Which number is even?

   5 ○   6 ○   7 ○   9 ○

1

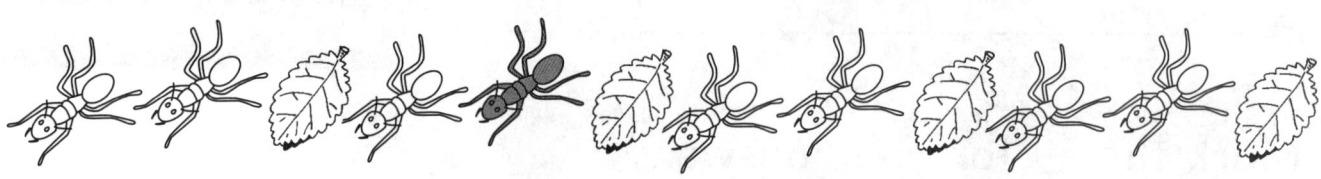

9. How many bugs is one fewer?

   5   6   7   9
   ○   ○   ○   ○

10. Which bug is shaded?

    ○ first   ○ seventh
    ○ fourth   ○ last

11. 9 is ___ 5.

    ○ greater than
    ○ equal to
    ○ less than
    ○ the same as

12. One more 🍃 makes the number ___.

    ○ zero   ○ odd
    ○ even   ○ eight

13. Which box is gray?

    ○ first   ○ fifth
    ○ third   ○ sixth

14. What number comes next?

    2, 3, 4, 2, 3, 4, ___.

    0   1   2   3
    ○   ○   ○   ○

15. Eleven is less than ___.

    ○ seven   ○ twelve
    ○ eight   ○ ten

16. A 🐜 ate 4 🍃.
    A 🐞 ate 1 fewer 🍃.
    How many 🍃 did the 🐞 eat?

    3   4   5   6
    ○   ○   ○   ○

2

Chapter 2 Test

Name _____

Mark the ○ for your answer.

1. What is the addition fact?

○ 3 + 1   ○ 4 + 0
○ 2 + 2   ○ 1 + 3

2. How many in all?

○ 0 + 1 = 3
○ 1 + 2 = 3
○ 2 + 1 = 3
○ 3 + 0 = 3

3. Add.

4 + 2 = ___

3   4   5   6
○   ○   ○   ○

4. Find the sum.

5 + 0 = ___

0   4   5   6
○   ○   ○   ○

5. Add.

2 + 2 = ___

2   3   4   5
○   ○   ○   ○

6. Add.

1¢ + 5¢ = ___

3¢   4¢   5¢   6¢
○   ○   ○   ○

Mark the ○ for your answer.

7. Add.

   4
   +1

   3 ○ 4 ○ 5 ○ 6 ○

8. Find the sum.

   3
   +0

   0 ○ 1 ○ 2 ○ 3 ○

9. Add.

   3
   +3

   3 ○ 4 ○ 5 ○ 6 ○

10. Find another sum of 6.

    5
    +1

    3+2 ○ 3+3 ○ 4+1 ○ 0+5 ○

11. Find the sum of the related facts.

    1   3
    +3   +1

    1 ○ 3 ○ 4 ○ 5 ○

12. Find the related fact for 2 + 4.

    ○ 4 + 2   ○ 4 + 4
    ○ 2 + 2   ○ 4 + 1

13. Show a sum of 5¢.

    2¢ + ? = 5¢

    1¢ ○ 2¢ ○ 3¢ ○ 4¢ ○

14. You are second in line. Two boys come. How many are in line now?

    3 ○ 4 ○ 5 ○ 6 ○

4

Name _____

Chapter 3 Test

Mark the ○ for your answer.

1. How many are left?

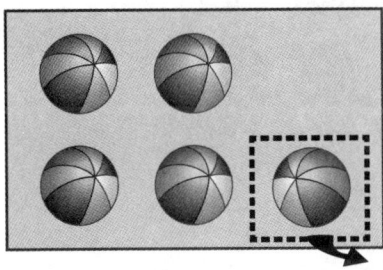

   1 ○   2 ○   3 ○   4 ○

2. How many are left?

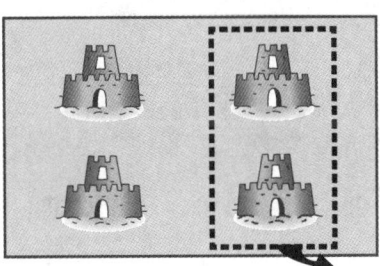

   1 ○   2 ○   3 ○   4 ○

3. Choose the subtraction sentence.

   ○ $3 - 1 = 2$ ○ $3 - 3 = 0$
   ○ $3 - 2 = 1$ ○ $3 - 0 = 3$

4. Find the difference.

   $6 - 4 = \_\_\_$

   2 ○   4 ○   5 ○   6 ○

5. 5  in all
   Take away 5.
   How many left?

   0 ○   1 ○   3 ○   5 ○

6. What comes next?

   $6 - 0 = 6$
   $5 - 0 = 5$
   $4 - 0 = \_\_\_$

   0 ○   4 ○   5 ○   6 ○

Mark the ○ for your answer.

7. Subtract.

$$\begin{array}{r}6¢\\-3¢\\\hline\end{array}$$

2¢ 3¢ 5¢ 6¢
○   ○   ○   ○

8. Find the difference.

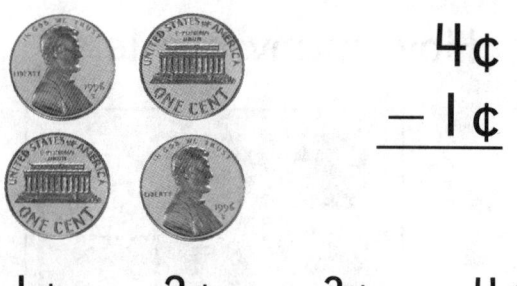
$$\begin{array}{r}4¢\\-1¢\\\hline\end{array}$$

1¢ 2¢ 3¢ 4¢
○   ○   ○   ○

9. Find the difference.

$$\begin{array}{r}2\\-2\\\hline\end{array}$$

0 1 2 4
○ ○ ○ ○

10. Find the difference.

$$\begin{array}{r}5\\-4\\\hline\end{array}$$

0 1 4 5
○ ○ ○ ○

11. Subtract.

$$\begin{array}{r}3\\-1\\\hline\end{array}$$

0 1 2 3
○ ○ ○ ○

12. Find the related addition fact.

$$\begin{array}{r}4\\-3\\\hline\\1\end{array}$$

○ 3 + 1
○ 2 + 2
○ 4 + 1
○ 0 + 4

13. Which tells the story?

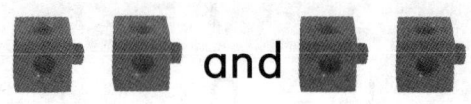

2 + 2 = 4   4 − 2 = 2
   ○   ○

14. Which tells the story?

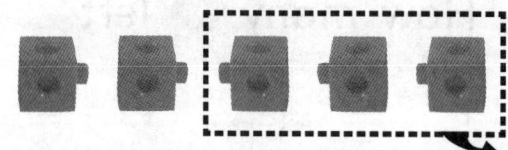

2 + 3 = 5   5 − 3 = 2
   ○   ○

Name _____

Chapter 4 Test

Mark the ○ for your answer.

1. Add.

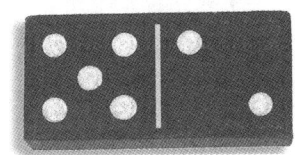

5 + 2 = ___

- 2 ○
- 5 ○
- 7 ○
- 8 ○

2. Find the sum.

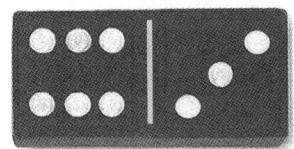

6 + 3 = ___

- 6 ○
- 8 ○
- 9 ○
- 10 ○

3. Add.

 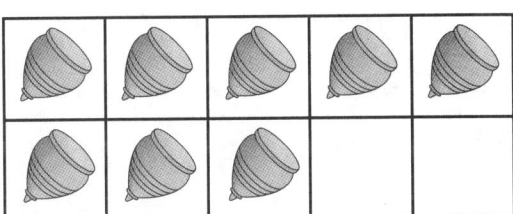

1 + 8 = ___

- 7 ○
- 8 ○
- 9 ○
- 10 ○

4. Add.

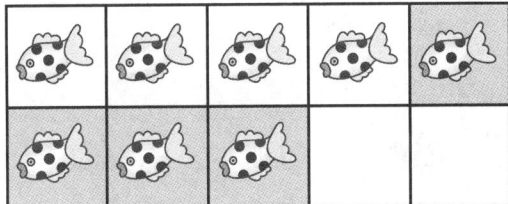

4 + 4 = ___

- 7 ○
- 8 ○
- 9 ○
- 10 ○

5. Find the sum.

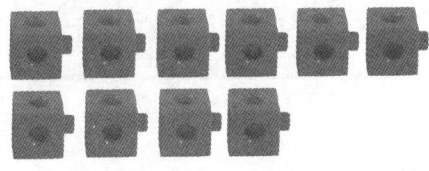

6
+4
___

- 9 ○
- 10 ○
- 11 ○
- 12 ○

6. Add.

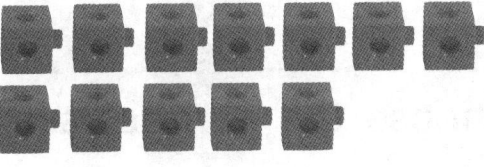

7
+5
___

- 9 ○
- 10 ○
- 11 ○
- 12 ○

Mark the ○ for your answer.

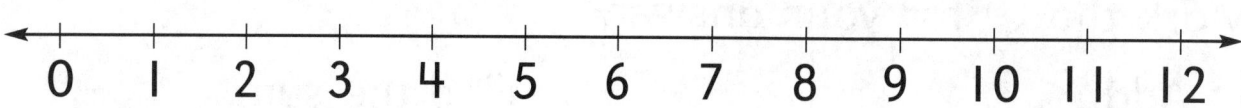

7. 🐸 jumps to 5.
   🐸 jumps 5 more.
   🐸 is on ___.

   7 8 9 10
   ○ ○ ○ ○

8. 🐸 can reach 11 with these hops.

   ○ 7 + 2   ○ 6 + 5
   ○ 9 + 1   ○ 8 + 0

9. What is another name for 9¢?

   ○ 8¢ + 1¢   ○ 7¢ + 3¢
   ○ 4¢ + 4¢   ○ 5¢ + 6¢

10. What is not another name for 12?

    ○ 7 + 5   ○ 9 + 2
    ○ 6 + 6   ○ 4 + 8

11. Add.

      3
      2
    +4
    ——

    ○ 9
    ○ 10
    ○ 11
    ○ 12

12. Add.

      2
      2
    +7
    ——

    ○ 9
    ○ 10
    ○ 11
    ○ 12

Choose the number sentence.

13. I had 8 ●.
    I found 3 more.

    8 + 3 = 11   8 − 5 = 3
        ○   ○

14. Pat had 6 ●.
    She lost 2 of them.

    6 + 2 = 8   6 − 2 = 4
        ○   ○

Chapter 5 Test

Name _____

Mark the ○ for your answer.

1. Subtract.

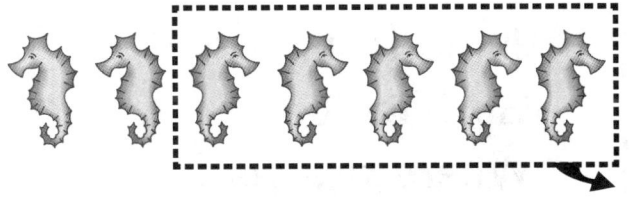

   7 − 5 = ___

   2 ○ 3 ○ 4 ○ 5 ○

2. Find the difference.

   8 − 4 = ___

   2 ○ 3 ○ 4 ○ 5 ○

3. Find the difference.

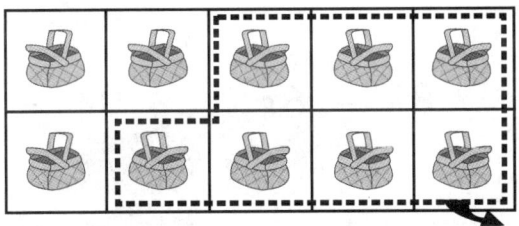

   10 − 7 = ___

   1 ○ 2 ○ 3 ○ 4 ○

4. Subtract.

   ○ 10 − 3   ○ 9 − 0
   ○ 6 − 3   ○ 9 − 3

5. Subtract.

   12 − 8 = ___

   0 ○ 4 ○ 7 ○ 8 ○

6. Find the difference.

   11 − 3 = ___

   5 ○ 6 ○ 7 ○ 8 ○

9

Mark the ○ for your answer.

```
←—+—+—+—+—+—+—+—+—+—+—+—+—→
  0  1  2  3  4  5  6  7  8  9 10 11 12
```

Find the difference. Use the number line.

7. Jump from 0 to 9.
   Go back 5.
   Where are you?

   2   3   4   5
   ○   ○   ○   ○

8. Jump from 0 to 10.
   Go back 6.
   Where are you?

   3   4   5   6
   ○   ○   ○   ○

9. Find the difference.

   $$\begin{array}{r} 7 \\ -0 \\ \hline \end{array}$$

   0   6   7   8
   ○   ○   ○   ○

10. What is another name for 5¢?

    ○ 8¢ – 6¢   ○ 9¢ – 3¢
    ○ 11¢ – 6¢   ○ 7¢ – 3¢

11. Which is not a name for 4?

    ○ 12 – 9   ○ 11 – 7
    ○ 7 – 3   ○ 10 – 6

12. Subtract.

    $$\begin{array}{r} 12¢ \\ -5¢ \\ \hline \end{array}$$

    5¢   7¢   8¢   10¢
    ○   ○   ○   ○

Choose the number sentence.

13. You see 8 🌹.
    You pick 2.

    8 + 2 = 10   8 – 2 = 6
    ○   ○

14. Jan made 5 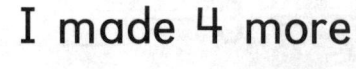.
    I made 4 more 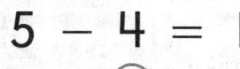.

    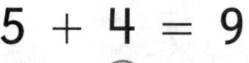 5 + 4 = 9   5 – 4 = 1
     ○   ○

Name _____

Chapter 6 Test

Mark the ○ for your answer.

1. How many?

   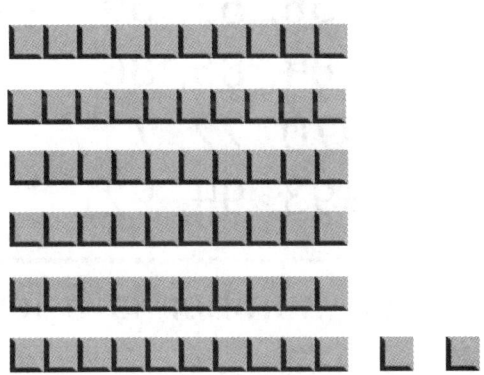

   7   25   52   62
   ○   ○   ○   ○

2. How many?

   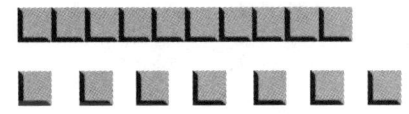

   ○ sixteen   ○ eighteen
   ○ seventeen   ○ nineteen

3. What is the number?

   3 tens 9 ones

   9   13   39   93
   ○   ○   ○   ○

4. What number is 10 more than?

   4 tens 0 ones

   10   40   50   60
   ○   ○   ○   ○

5. How many groups of 5?

   2   3   5   15
   ○   ○   ○   ○

6. What number comes just before?

   ___, 22

   23   21   30   20
   ○   ○   ○   ○

11

Mark the ○ for your answer.

7. What number comes just after?

69, ___

- 80 ○
- 7 ○
- 70 ○
- 68 ○

8. Which numbers are in order?

- ○ 56, 57, 59
- ○ 84, 85, 86
- ○ 70, 72, 73
- ○ 93, 94, 92

9. Which number is not greater than 38?

- 58 ○
- 48 ○
- 28 ○
- 83 ○

10. Which number is 5 less than 95?

- 80 ○
- 85 ○
- 90 ○
- 100 ○

11. Count by 2.

42, 44, 46, ___

- 47 ○
- 48 ○
- 49 ○
- 50 ○

12. Count by 5.

20, 25, ___, 35

- 30 ○
- 35 ○
- 40 ○
- 45 ○

13. I am an odd number between 50 and 60. You say me when you count by 5.

- 52 ○
- 55 ○
- 56 ○
- 58 ○

14. I am a number you say when you count by 10. I am greater than 70.

- 60 ○
- 75 ○
- 80 ○
- 85 ○

Name _____

Mid-year Test

Mark the ○ for your answer.

1. How many ●?

○ 7   ○ 8   ○ 9   ○ 10

2. How many 🍎?

🍎🍎🍎 and 🍎🍎

○ 2   ○ 3   ○ 4   ○ 5

3. Find the sum.

← 5 6 7 8 9 10 11 12 →

7 + 5

○ 9   ○ 11   ○ 12   ○ 13

4. What number is missing?

← 5 6 7 ☐ 9 10 11 12 →

○ 6   ○ 8   ○ 10   ○ 12

5. What number is just before 7?

○ 4   ○ 5   ○ 6   ○ 8

6. 🐢 🐢 🐢 ✗🐢 🐢

The turtle with the ✗ is _____.

○ first   ○ third
○ second   ○ fourth

7. Find the sum.

  4   1
+1   +4
___   ___

○ 1   ○ 3   ○ 4   ○ 5

8. Find another sum of 5.

3 + 2 = 5

○ 0 + 4   ○ 3 + 4
○ 1 + 4   ○ 3 + 3

13

Mark the ○ for your answer.

9. What is 1 fewer?

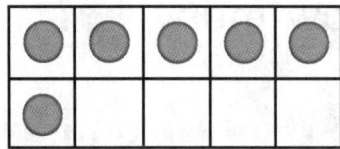

   3   4   5   6
   ○   ○   ○   ○

10. What is the addition fact?

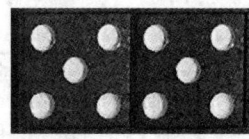

○ 10 − 5   ○ 5 + 4
○ 6 + 4   ○ 5 + 5

11. What is not a name for 10?

○ 5 + 5   ○ 4 + 7
○ 12 − 2 ○ 3 + 5 + 2

12.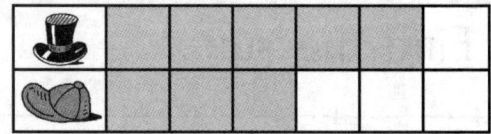

How many hats in all?
○ 6 + 2 = 8
○ 4 + 4 = 8
○ 5 + 3 = 8

13. Find the sum.

    3
    4
 +4
 ―――

 9   10   11   12
 ○   ○   ○   ○

14. Find the sum.

 5   7   8   10
 ○   ○   ○   ○

15. The sixth 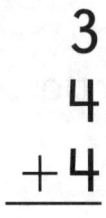 in a pattern is red. There are 3 more 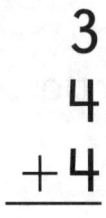. How many 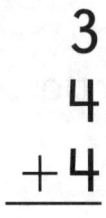 in all?

 5   7   9   11
 ○ ○ ○ ○

16. Jon has 7 . He gets 2¢. Then he finds 3¢. How many in all?

 12¢ 11¢ 10¢ 9¢
  ○   ○   ○   ○

Mid-year Test

Name _____

Draw or model to compare.

17. Find the number sentence.

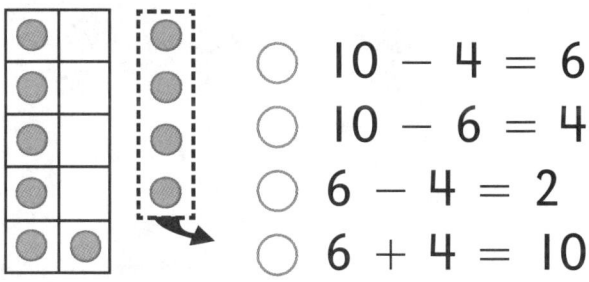

○ 10 − 4 = 6
○ 10 − 6 = 4
○ 6 − 4 = 2
○ 6 + 4 = 10

18. Subtract.

10 − 3 = ___

13   7   8   6
○   ○   ○   ○

19. Find the difference.

11¢
− 6¢

2¢   3¢   4¢   5¢
○   ○   ○   ○

20. Find the missing number.

9 − 9 = ☐

6 + ☐ = 6

0   1   4   12
○   ○   ○   ○

21. Find the missing number.

12 − ☐ = 4

7   8   9   10
○   ○   ○   ○

22. Find the difference of 2.

11 − 9 = 9 − ☐

5   7   6   8
○   ○   ○   ○

23. Jose has 12¢. Juan has 8¢. How much more does Jose have?

4¢   2¢   5¢   3¢
○   ○   ○   ○

24. You draw a dozen 🎈. You color 7 green. How many are not green?

5   6   7   4
○   ○   ○   ○

Mark the ◯ for your answer.

25. What is the number?

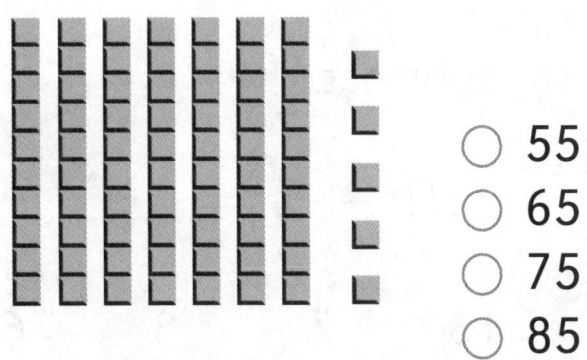

- ◯ 55
- ◯ 65
- ◯ 75
- ◯ 85

26. How many in each group?

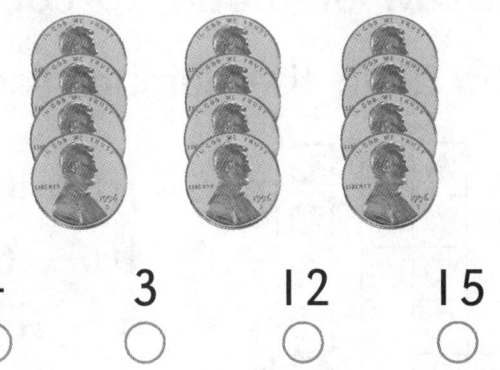

4   3   12   15
◯   ◯   ◯   ◯

27. What is the missing number?

50, 60, 70, ___, 90

75   85   80   95
◯   ◯   ◯   ◯

28. What is the number?

eighty-seven

78   70   80   87
◯   ◯   ◯   ◯

29. Mark the number.

8 tens 7 ones

78   83   87   89
◯   ◯   ◯   ◯

30. Which number is the greatest?

◯   ◯   ◯   ◯

31. Mark the missing number.

32, 34, 36, ___, 40

37   28   38   39
◯   ◯   ◯   ◯

32. I am between 70 and 80. I am a number you say when you count by 5.

72   74   75   78
◯   ◯   ◯   ◯

Chapter 7 Test

Name _____

Mark the ◯ for your answer.

1. How much?

30¢ 40¢ 50¢ 60¢
◯   ◯   ◯   ◯

2. How much?

3¢   11¢ 15¢ 30¢
◯   ◯   ◯   ◯

3. How much?

20¢ 22¢ 31¢ 40¢
◯   ◯   ◯   ◯

4. How much more to trade for a quarter?

4¢   5¢   7¢   9¢
◯   ◯   ◯   ◯

5. Which amount is between 30¢ and 40¢?

◯   ◯   ◯   ◯

6. How much?

42¢ 47¢ 50¢ 55¢
◯   ◯   ◯   ◯

Mark the ○ for your answer.

7. What time is it?

- ○ 6:00
- ○ 6:30
- ○ 7:00
- ○ 7:30

8. Which shows eleven thirty?

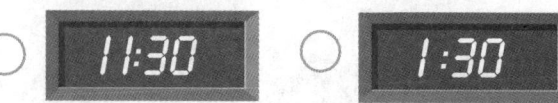

9. What time is it?

- ○ 1 o'clock
- ○ 2 o'clock
- ○ half past 2
- ○ 30 minutes after 6

10. What time is 1 hour from now?

- ○ eleven
- ○ 11 thirty
- ○ twelve thirty
- ○ 1 o'clock

11. When will they be done?

 starts

12:00   2:00   12:30   11:30
  ○   ○   ○   ○

12. When will it be over?

 starts

1:00   2:30   12:30   5:00
 ○   ○   ○   ○

🌸🌸🌸🌸	April	🌸🌸🌸🌸				
S	M	T	W	T	F	S
	1	2	3	4	5	6
7	8	9	10	11	12	13

13. One week from April 6 is April ____.

10   12   13   15
 ○   ○   ○   ○

14. The first Wednesday is April ____.

1   3   8   10
○   ○   ○   ○

Name _____

Chapter 8 Test

Mark the ○ for your answer.

1. Which is a square?

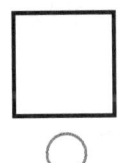

2. Which is a solid?

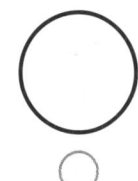 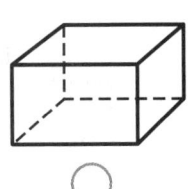

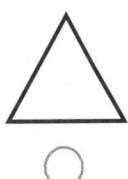 

3. How many sides and corners?

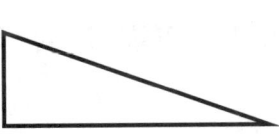

0   1   2   3
○   ○   ○   ○

4. Which has more sides than corners?

 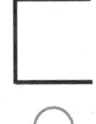
○   ○

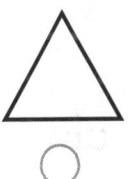 
○   ○

5. Which does not have equal parts?

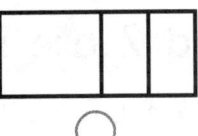 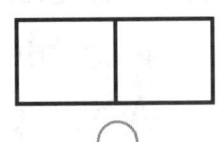
○   ○

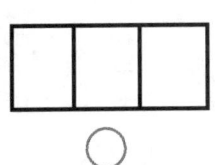 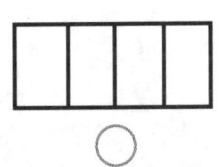
○   ○

6. Which has the same size and shape as ▭ ?

 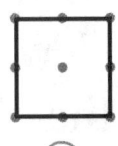
○   ○

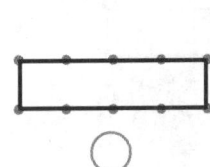 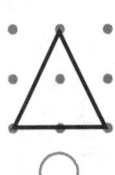
○   ○

19

Mark the ○ for your answer.

7. Which can not stack?

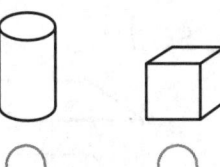 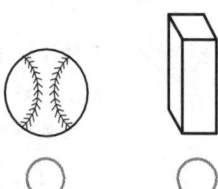 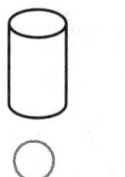 
○ ○ ○ ○

8. Which can not roll?

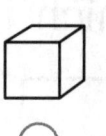 
○ ○ ○ ○

9. Which is a pyramid?

 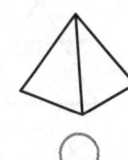
○ ○

 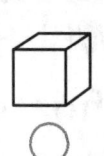
○ ○

10. Which shows $\frac{1}{3}$ shaded?

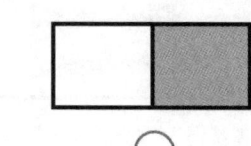 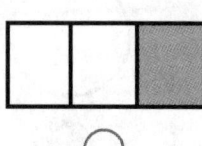
○ ○

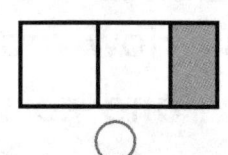
○ ○

11. How often will you spin a 1?

○ always
○ sometimes
○ never

12. Which shows $\frac{1}{2}$ shaded?

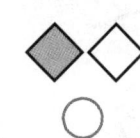 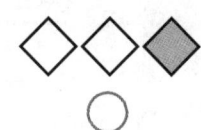
○ ○

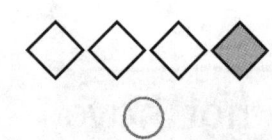 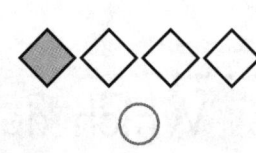
○ ○

13. Solve. How many ▫ in the shape?

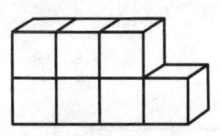

4   5   6   7
○   ○   ○   ○

14. You stack 4 blocks. 1 is blue and 2 are red. What part is not red or blue?

$\frac{1}{2}$ $\frac{1}{3}$ $\frac{1}{4}$ $\frac{1}{6}$
○   ○   ○   ○

Chapter 9 Test

Name _____

Mark the ○ for your answer.

1. Add.

   4 tens + 3 tens = ____

   | 40 | 43 | 70 | 73 |
   | ○ | ○ | ○ | ○ |

2. Add.

   20
   +22
   ___

   | 40 | 22 | 24 | 42 |
   | ○ | ○ | ○ | ○ |

3. Subtract.

   6 tens − 5 tens = ____

   | 10 | 40 | 65 | 1 |
   | ○ | ○ | ○ | ○ |

4. Subtract.

   80¢
   −30¢
   ____

   | 5¢ | 30¢ | 40¢ | 50¢ |
   | ○ | ○ | ○ | ○ |

5. Add.

   35
   +21
   ___

   | 55 | 56 | 65 | 66 |
   | ○ | ○ | ○ | ○ |

6. Find the sum.

   50
   +47
   ___

   | 79 | 87 | 90 | 97 |
   | ○ | ○ | ○ | ○ |

7. Find the difference.

   87 − 40 = ____

   | 30 | 37 | 40 | 47 |
   | ○ | ○ | ○ | ○ |

8. Subtract.

   8 dimes
   −5 dimes
   _____

   | 20¢ | 30¢ | 40¢ | 50¢ |
   | ○ | ○ | ○ | ○ |

Mark the ○ for your answer.

9. Add.

$$\begin{array}{r} 23¢ \\ +65¢ \\ \hline \end{array}$$

68¢ ○   78¢ ○   88¢ ○   98¢ ○

10. Add.

$$\begin{array}{r} 71 \\ +4 \\ \hline \end{array}$$

57 ○   75 ○   76 ○   85 ○

11. Subtract.

$$\begin{array}{r} 47 \\ -13 \\ \hline \end{array}$$

23 ○   34 ○   43 ○   44 ○

12. Find the difference.

$$\begin{array}{r} 69 \\ -42 \\ \hline \end{array}$$

27 ○   36 ○   37 ○   72 ○

13. Add.

$$\begin{array}{r} 58¢ \\ +27¢ \\ \hline \end{array}$$

75¢ ○   71¢ ○   85¢ ○   81¢ ○

14. Add.

$$\begin{array}{r} 32¢ \\ +49¢ \\ \hline \end{array}$$

71¢ ○   77¢ ○   81¢ ○   87¢ ○

15. Solve.

Max has 32 . He buys 19 more . About how many  in all?

20 ○   30 ○   40 ○   50 ○

16. Solve.

Has 21¢. Has 3 coins.

○   ○   ○   ○

Name _____

Chapter 10 Test

Mark the ○ for your answer.

1. How many 🖇 around?

   4   5   6   7
   ○   ○   ○   ○

2. How many ☐ around?

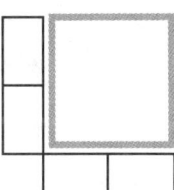

   2   4   6   8
   ○   ○   ○   ○

3. Measure. How many inches?

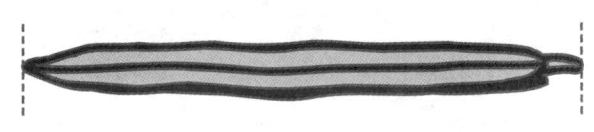

   1   2   3   4
   ○   ○   ○   ○

4. Measure. How many inches?

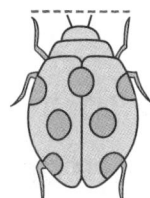

   1   2   3   4
   ○   ○   ○   ○

5.

   ○ less than 1 inch
   ○ less than 1 foot
   ○ more than 1 foot

6.

   ○ less than 1 inch
   ○ less than 1 foot
   ○ more than 1 foot

7.

   1 cup   1 pint   1 quart
   ○   ○   ○

8.

   ○ less than 1 pound
   ○ 1 pound
   ○ more than 1 pound

23

Mark the ◯ for your answer.

9. Measure. How high?
   - ◯ 1 centimeter
   - ◯ 4 centimeters
   - ◯ 7 centimeters

10. About how much?
    - ◯ less than 1 liter
    - ◯ 1 liter
    - ◯ more than 1 liter

11. About how heavy?
    - ◯ less than 1 kilogram
    - ◯ 1 kilogram
    - ◯ more than 1 kilogram

12. The temperature is ____.
    - 20°F ◯
    - 50°F ◯
    - 80°F ◯

What is the best measuring tool to use?

13. How long?
    - ◯ ruler
    - ◯ cup
    - ◯ scale

14. How heavy?
    - ◯ ruler
    - ◯ cup
    - ◯ scale

15. ___ units from 🏠 to 🌊
    - 6 ◯
    - 7 ◯
    - 9 ◯
    - 10 ◯

16. Tia drank a cup of juice. Gina drank a liter. Mia drank a pint. Who drank the most juice?
    - Tia ◯
    - Gina ◯
    - Mia ◯

24

Name _____

Chapter 11 Test

Mark the ◯ for your answer.

1. Add.

    6   7
   +7 +6

   13 14 15 16
   ◯   ◯   ◯   ◯

2. Add.

    8   6
   +6 +8

   14 15 16 17
   ◯   ◯   ◯   ◯

3. Find the missing number.

   9 + ☐ = 14

   14 − ☐ = 9

   4 5 6 7
   ◯ ◯ ◯ ◯

4. Add.

    9¢
   +9¢

   15¢ 16¢ 17¢ 18¢
   ◯   ◯   ◯   ◯

5. Find the missing number.

   7 + 9 = 8 + ☐

   5 6 7 8
   ◯ ◯ ◯ ◯

6. 8 + 7 and 6 + 9 are names for

   15 16 17 18
   ◯   ◯   ◯   ◯

7. Which is not a name for 13?

   ◯ 5 + 8   ◯ 6 + 5
   ◯ 9 + 4   ◯ 7 + 6

8. 7 + 8 + 3 = ___

   15 16 17 18
   ◯   ◯   ◯   ◯

25

Mark the ◯ for your answer.

9. Subtract.

16
− 7

◯ 6
◯ 7
◯ 8
◯ 9

10. Find the difference.

13
− 9

◯ 4
◯ 5
◯ 6
◯ 7

11. Find the missing number.

6 + 9 = ☐

9 + 6 = ☐

☐ − 9 = 6

☐ − 6 = 9

14 15 16 17
◯   ◯   ◯   ◯

12. Find the missing number.

18 − ☐ = 9

7   8   9   10
◯   ◯   ◯   ◯

13. Find the difference.

15¢
− 6¢

◯ 9¢
◯ 10¢
◯ 11¢
◯ 12¢

14. Find the difference.

17¢
− 8¢

◯ 5¢
◯ 7¢
◯ 8¢
◯ 9¢

15. Ann saw 5 🐥.
Mary saw 9 🐥.
2 🐥 were white.
How many 🐥 in all?

12 14 16 18
◯   ◯   ◯   ◯

16. Tim had 13 🐟.
Joe had 4 🐟.
How many more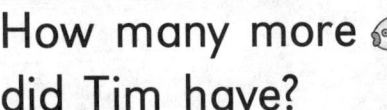
did Tim have?

5   6   8   9
◯   ◯   ◯   ◯

26

Name _____

End-of-Year Test

Mark the ○ for your answer.

1. What number is missing?

   ___, 43, 44, 45

   46 41 42 40
   ○   ○   ○   ○

2. What comes between?

   66, ___, 68

   64 65 67 69
   ○   ○   ○   ○

3. Which is less than 53?

   49 54 61 62
   ○   ○   ○   ○

4. 7 tens 4 ones equals _?_

   44 47 70 74
   ○   ○   ○   ○

5. How many?

   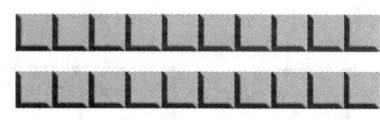

   14 24 25 26
   ○   ○   ○   ○

6. What comes next?

   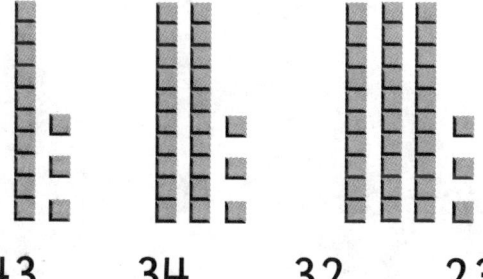

   43 34 32 23
   ○   ○   ○   ○

7. Which is the fair trade?

   ○   ○   ○   ○

8. What time is it?

   ○ half past 6   ○ 3 o'clock
   ○ three thirty   ○ 6 o'clock

27

Mark the ○ for your answer.

9. 60 − 30 = ___

90   50   40   30
○   ○   ○   ○

10. 56 + 30 = ___

26   46   66   86
○   ○   ○   ○

11. Find the sum.

  42¢
 +35¢
 ─────

77¢   67¢   57¢   17¢
○   ○   ○   ○

12. Find the difference.

  78¢
 −54¢
 ─────

20¢   34¢   14¢   24¢
○   ○   ○   ○

13. Find the missing number.

7 + 3 + 2 = 6 + 2 + ___

3   4   5   6
○   ○   ○   ○

14. Which shows equal parts?

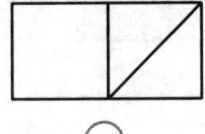 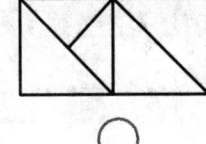
○   ○

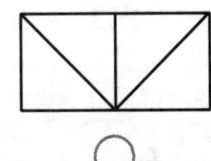 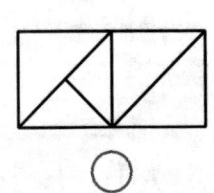
○   ○

15. What part is shaded?

$\frac{1}{2}$ $\frac{1}{3}$ $\frac{1}{4}$ $\frac{2}{3}$
○   ○   ○   ○

16. Pam has 5 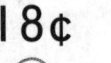 and 8 . She buys this toy. How much money is left?

18¢   28¢   20¢   10¢
○   ○   ○   ○

Name _____

End-of-Year Test

Mark the ○ for your answer.

17. Which fact does not belong to this family?

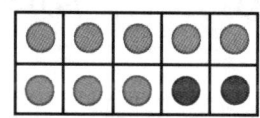

○ 8 + 2 = 10
○ 2 + 8 = 10
○ 7 + 3 = 10
○ 10 − 2 = 8

18. A 🏷 costs 53¢. You have 2  and 3 . What coins do you still need?

○   ○   ○   ○

19. About how many?

15   20   24   25
○   ○   ○   ○

20. Close the figure. The perimeter is about

____ □.

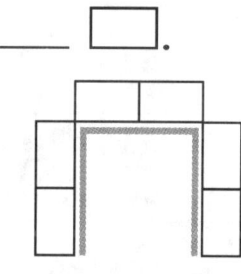

○ 2
○ 6
○ 8
○ 10

21. You are less likely to pick ● than ● from ____.

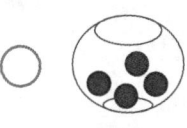   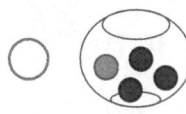
○   ○

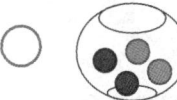   
○   ○

22. Which holds the least?

○   ○   ○

23. Which piece of yarn is about 4 inches?

29

Mark the ○ for your answer.

	May					
S	M	T	W	T	F	S
		1	2	3	4	5
6	7	8	9	10	11	12
13	14	15	16	17	18	19
20	21	22	23	24	25	26
27	28	29	30	31		

24. May 1 is on ____ .
   ○ Monday   ○ Wednesday
   ○ Tuesday ○ Thursday

25. There are ____ Fridays in May.

   3   4   5   6
   ○   ○   ○   ○

26. ____ has the most money.

   Jon   Jim   Jill
   ○   ○   ○

27. ____ has 35¢.

   Jon   Jim   Jill
   ○   ○   ○

28. ____ have 25¢ each.

   ○ Jill and Jim
   ○ Jane and Jill
   ○ Jim and Jill

29. How much money does Jane need to have as much as Jon?

   10¢   20¢   25¢
   ○   ○   ○

30

Chapter 12 Test

Mark the ○ for your answer.

1. Write the number sentence.

   3 + 7 = 50 − ___

   10 ○   20 ○   30 ○   40 ○

2. Write the number sentence.

   19 − 6 = 8 + ___

   3 ○   4 ○   5 ○   6 ○

3. Use + or −.

   15 ○ 4 ○ 6 = 5

   + ○   − ○   + and − ○

4. Find the missing number.

   ▲ = 7   ■ = 4

   ▲ + ▲ − ■ = ___

   8 ○   10 ○   11 ○   12 ○

5. Find the sum.

   47
   +35

   28 ○   72 ○   81 ○   82 ○

6. Add.

   69¢
   + 7¢

   66¢ ○   67¢ ○   76¢ ○   77¢ ○

7. Subtract.

   42¢
   −18¢

   24¢ ○   26¢ ○   34¢ ○   42¢ ○

8. Find the difference.

   90
   −63

   20 ○   26 ○   27 ○   37 ○

Mark the ○ for your answer.

9. Find the sum.

```
  13     ○ 77
  24     ○ 78
 +41     ○ 87
         ○ 88
```

10. Find the sum.

```
  40     ○ 88
  32     ○ 89
 +17     ○ 98
         ○ 99
```

11. Find the sum.

```
 326     ○ 114
+212     ○ 538
         ○ 539
         ○ 548
```

12. Find the difference.

```
 654     ○ 320
-234     ○ 310
         ○ 410
         ○ 420
```

13. What is the number?

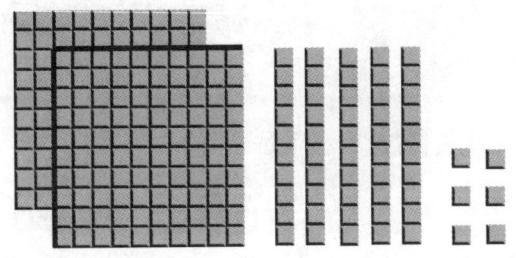

| 165 | 156 | 256 | 265 |
| ○ | ○ | ○ | ○ |

14. Brendon has 4 dimes and 15 pennies. How much money does he have?

| 19¢ | 45¢ | 54¢ | 55¢ |
| ○ | ○ | ○ | ○ |

15. How many groups of 2?

| 1 | 2 | 4 | 8 |
| ○ | ○ | ○ | ○ |

16.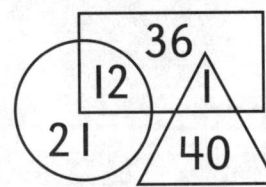

What is the sum of the numbers in the rectangle?

| 33 | 37 | 41 | 49 |
| ○ | ○ | ○ | ○ |

32